SPACE GUIDES

EXPLORING THE EARTH

PETER GREGO

QED Publishing

QED

First published in the UK in 2007 by
QED Publishing
A Quarto Group company
226 City Road
London EC1V 2TT
www.qed-publishing.co.uk

Reprinted in 2008

A catalogue record for this book is available from the British Library.

ISBN 978 1 84835 014 4

Written by Peter Grego
Produced by Calcium
Editor Sarah Eason
Illustrations by Geoff Ward
Picture Researcher Maria Joannou

Publisher Steve Evans
Creative Director Zeta Davies
Senior Editor Hannah Ray

Printed and bound in China

Picture credits

Key: T = top, B = bottom, C = centre, L = left, R = right, FC = front cover, BC = back cover

Corbis/Bettmann 16, /Tom Bean 24–25, /Jonathan Blair 22–23, /Ric Ergenbright 26–27, /David Muench 23B; **ESA**/NASA and G. Bacon 6; **Getty Images**/Iconica FCB, 11B, /Imagebank 20–21, /Photodisc FCT, 4, BC, /Taxi 16–17; **NASA** 9B, 10, 18, 19B, /GSFC 1, 19T, /GSFC/Jacques Descloitres/MODIS Rapid Response Team 12–13, /GSFC/Craig Mayhew and Robert Simmon 28–29, /IKONOS 3, /JPL-Caltech 7T, 7B, 8, /USGS 14, 15T; **NGDC** 29; **NOAA** 27, 28, /Lieutenant Philip Hall 13; **Science Photo Library**/Chris Butler 26, /Mark Garlick 5T, /David A Hardy 9T, /Walter Pacholka/Astropics 11T, / Soames Summerhays FCC, /Dirk Wiersma 23T; **USGS** 24, 25, /Seth Moran 21

Words in **bold** can be found in the Glossary on pages 30–31.

Website information is correct at time of going to press. However, the publishers cannot accept liability for any information or links found on third-party websites.

Contents

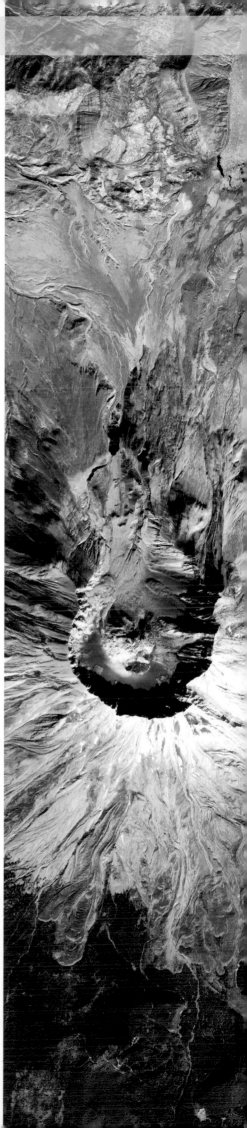

The blue planet

Our **planet**, the Earth, is one of eight planets that make up our **Solar System**. They circle a star called the **Sun**. We depend upon the Sun's tremendous heat and light to create our weather and help keep us, other animals and plants alive.

The Earth is a medium-sized planet, the fifth largest in the Solar System. It measures 12 756km in diameter, so it would take you several years to walk around it.

Our place in the Universe

Earth is our home in the vast **Universe**. Since the first astronaut blasted off into **space**, in April 1961, fewer than 1000 people have managed to see our planet from space. There, it appears a beautiful, blue globe spinning through the inky black vastness of space. It is an overwhelming sight.

Our planet, the Earth, photographed by ⇑ astronauts returning from the **Moon**.

Mercury

Jupiter

Uranus

Earth

Sun

Venus

Mars

Saturn

Neptune

⬆ The eight planets of our Solar System, shown to scale. The edge of the Sun – a globe so big that a million Earths could fit inside it – is shown on the far left.

Precious world

From space, borders between countries and peoples cannot be seen on Earth. It reminds us that our planet is just one world, a fragile place that we must do our best to protect.

Key Concept

Our neighbours

The Earth is one of eight major planets orbiting **the Sun. The four planets nearest the Sun are solid and rocky, like the Earth. The four planets furthest from the Sun are balls of gas, with no solid surface.**

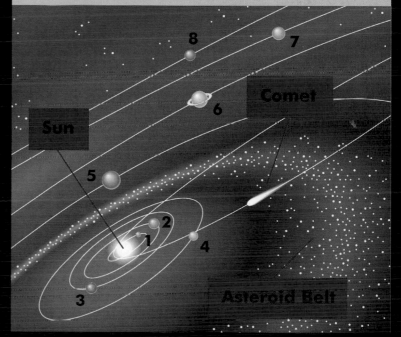

8

7

Comet

6

Sun

5

1

2

4

3

Asteroid Belt

This diagram shows the eight planets of our Solar System in orbit around the Sun. (1. Mercury, 2. Venus, 3. Earth, 4. Mars, 5. Jupiter, 6. Saturn, 7. Uranus, 8. Neptune.) The Solar System also includes the **Asteroid Belt** and many **comets**.

⬆

Birth of the Solar System

More than 4.6 billion years ago, the blast from an exploding star disturbed a nearby cloud of dust and gas. The blast pushed parts of the cloud together, which made them denser (thicker and bulkier). These areas had more **gravity** – a force that pulls little objects towards bigger objects. Gravity made the denser parts of the cloud gather together into more solid shapes. Scientists think this is how the material that makes up the Sun was first gathered together.

A star is born

As the dust cloud became heavier, it began to spin. This produced a flattened disk of dust and gas, which spun around the denser material at its centre. As the centre contracted, it became even thicker and hotter. Eventually, the centre became so hot that it began to burn. It became a star – the newborn Sun had begun to shine.

⬆ The newborn Earth was a hot globe baking in the heat of the young Sun.

The young Sun was surrounded by clouds of dust and gas, which clumped together to create the planets.

The young Solar System

Meanwhile, parts of the disk of dust and gas surrounding the Sun were also drawn together by gravity. This produced some large planets, lots of moons, and thousands of rocky **asteroids** and icy comets. All of these objects made up the young Solar System.

Powerful winds from the new Sun blew away all the light gases in the inner Solar System. Only objects made out of heavier materials remained. Our Earth was one of these objects.

The gravity of the ⇨ young planets swept up any rock and dust that orbited them, but larger chunks of rocks remained further out in the solar system. These made up the Asteroid Belt between Mars and Jupiter.

Amazing

Ancient zircon

Crystals taken from rocks in Australia were found to be 4.4 billion years old. They formed just after the Earth was created.

The young Earth

As asteroids smashed into its surface, the young Earth's temperature rose. It's gravity kept pulling in more material and, as the Earth grew bigger, its inside continued to get hotter and hotter. It soon became hot enough to melt the metals in its rocks. These **molten** metals sank to the Earth's centre and formed its **core**. Lighter material rose to form the planet's outer layer, called its crust. Between the core and the crust is a layer of very hot molten rock, called the **mantle**.

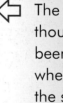

The Moon is thought to have been formed when a planet the size of Mars crashed into the Earth.

A big whack

The Moon was formed by a crash. Scientists believe that a planet half the size of ours struck the young Earth, throwing out a massive sheet of melted material. Much of this material was pulled back together by gravity, and formed the Moon.

Amazing

Asteroid attack

For more than a billion years after the Earth was formed, asteroids frequently smashed through its thin, rocky crust. This allowed hot, melted rock to burst through and spread over the surface as sheets of bubbling **lava**. Over time, these helped to build up the Earth's crust.

The Moon was also hit by asteroids, which formed **craters** on its surface. If the Earth's surface hadn't continued to be changed by the movements of its crust, **volcanoes** and **weathering**, it would have as many craters as the Moon.

Both the Earth and the Moon were hit by asteroids in their early history.

Lakes and seas

Once asteroids started hitting the planet less often, the Earth's crust began to cool and thicken. Volcanoes continued to erupt, spewing out lava and releasing water vapour into the air. As the Earth cooled, water vapour turned into liquid. Small puddles slowly turned into lakes and seas. Icy comets sometimes hit the Earth, too, adding to the water on its surface when they melted. The Earth now had areas of land and of water.

We can see craters billions of years old on the surface of the Moon. Despite being very old, they look quite new. They have steep walls and sharp rims, and large mountains often rise from their centres.

Continents, plates and mountains

All the landmasses (**continents**) seen on Earth today were once joined together in a single, giant landmass known as Pangaea, which was surrounded by a huge ocean. However, the molten rock of the mantle pushed up on the Earth's crust, causing it to break up into smaller pieces, called **plates**. And as the crust broke into plates, Pangaea was split into smaller pieces, too.

Shifting plates

Gaps between the plates became wider as molten **magma** from the mantle pushed into them. This made South America break away from Africa, and India and Australia break away from Antarctica. A hundred million years ago, molten magma also separated Europe from North America. Ocean water filled the space between these new continents, creating new seas and oceans.

← The red lines show the plates that make up the Earth's crust.

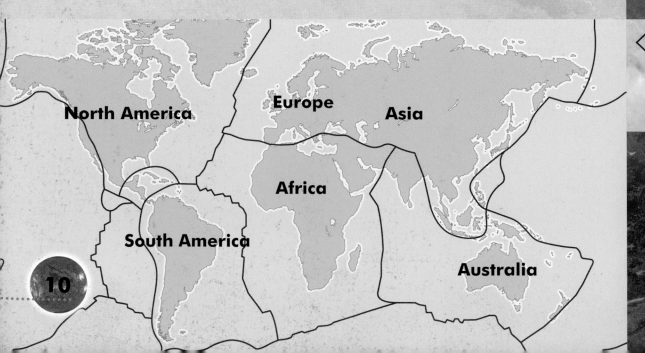

North America

Europe

Asia

Africa

South America

Australia

Alfred Wegener (1880–1930)

We know that the plates of the Earth are moving, and that the Earth used to look very different. But how did we find this out? By looking at the shapes of the continents, we can see that they would once have fitted together, like a giant jigsaw. For instance, if the west coast of Africa was placed next to the east coast of South America, they would fit snugly together. Their rocks and fossils are also similar, suggesting they were once one landmass. The theory that continents move, or drift, was first suggested by German scientist Alfred Wegener, only 100 years ago. Although they move by just a few centimetres each year, over millions of years this adds up to a great distance. We call this moving of continents continental drift.

⇧ The Himalayas is the Earth's biggest mountain range – and it's still growing. It was created when India pushed into southern Asia, forcing up the Earth's crust.

Mountains

Mountains are formed when two plates push against each other, forcing up Earth's crust. Sometimes the crust beneath an ocean pushes against the crust beneath a continent. When this happens, the crust beneath the continent is pushed up to make a mountain. The biggest mountain ranges, however, occur when the plates beneath two continents collide and both crumple up.

⇧ Mountains are one of the results of plate movement.

Earth changes

If the Earth could have been seen from space 100 years ago, the shapes and positions of the main areas of land would look the same as those seen today. However, the Earth's surface is constantly changing. Some changes happen quickly, such as those caused by earthquakes and volcanoes. Others happen over a longer period of time, such as the wearing away of rocks by weather and water, or continental drift.

Parts of Earth's surface are built up when material gathers over a long period of time. This is called sedimentation. The small islands at the mouth of the River Lena, in Russia, were formed in this way. They were made when mud, small bits of rock and other loose material flowed downstream and collected where the river meets the sea. The islands break up the river into a network of smaller channels, called a delta.

Land shifts

We humans have only been recording our planet for a few thousand years. This is an incredibly short time compared to the life of the Earth. The Earth is about 4.6 billion years old. If we could hop into a time machine and zoom through the history of the Earth so fast that each century passed in one second, we would be amazed to see mountains being built and wearing away, spaces for oceans being made or squeezed out of existence, coastlines being created and destroyed, and new continents being formed. At the same time, the sea level would rise and fall over time, as the **ice caps** at the north and south poles grew or shrank.

Key Concept

Continental drift

Powered by deep movements within the Earth's hot mantle, continental drift is still happening. The Atlantic Ocean is widening by about 4cm each year. India is continuing to collide with south Asia, crumpling the solid crust between the two continents. This is forcing up the mighty Himalayan Mountains even further.

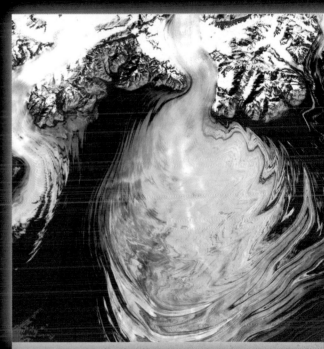

North Sea

Baltic Sea

antic ean

Black Sea

Mediterranean Sea

⬆ This glacier (part of an ice cap) is slowly melting, as the Earth's **climate** warms up. In a few hundred years' time the ice may have melted completely. If many glaciers melt, sea levels will rise and flood large areas of land. Earth will look different once again.

⬅ This map shows how Europe might look if the sea level rises by 100m. The lighter blue areas show land that would be flooded.

Volcanoes and earthquakes

Volcanoes are formed in places where hot, molten rock breaks through the Earth's crust and onto the surface. Like mountains, volcanoes are usually found where plates collide or move apart. There are about 1000 active volcanoes around the world. No two volcanoes are exactly alike. Volcanoes around the edges of continents are often tall and steep-sided, built up over the years by eruptions of thick, slow-moving lava and piles of ash.

Vesuvius

One of the most famous volcanoes is Vesuvius, on the west coast of Italy. Vesuvius began as a small, volcanic hill about 25 000 years ago, and it is now 1300m high. Almost 2000 years ago, an eruption of Vesuvius destroyed the nearby Roman towns of Pompeii and Herculaneum.

In 79CE, the ancient Roman towns of Pompeii and Herculaneum were completely buried beneath volcanic ash and rock when the volcano Vesuvius erupted. The bodies of victims (like those shown here) were preserved in layers of volcanic ash and can still be seen today.

This is Mount St Helens. A mighty mountain peak used to be where the gaping crater is now.

Amazing

Mount St Helens

One of the most devastating eruptions of recent times happened in May 1980 in Washington State, north-western USA, when Mount St Helens blew its top. The immense explosion blasted a column of ash into the air up to 25km high.

Earthquakes

Earthquakes happen when two plates are pushed together and suddenly slide past each other for a short distance.

Earthquakes range from mild vibrations to devastating shakings of the ground, which can topple buildings. Small earthquakes happen in Britain each year, but few cause any real damage. Large buildings in areas that suffer from large earthquakes, such as California and Japan, are built to withstand the violent jolting of the ground. Buildings in poorer countries may not be so well made, and great loss of life can occur when an earthquake happens near a big city in which many people live.

Asia

North America

Pacific Ocean

Australia

South America

☐ **Ring of Fire**

⇧ Earthquakes are common in an area known as the Pacific Ring of Fire. Here, the plate beneath the ocean is being pushed beneath the surrounding continental plates.

Inside the Earth

The Earth's natural movements have revealed a lot about the rocks that lie beneath its surface. Layers of rock that were once buried have been lifted up as the crust has moved and shifted. When the Grand Canyon formed, in Arizona in the USA, it made a cut in the Earth's crust 1600m deep, uncovering rocks that are more than a billion years old!

Drilling into Earth

Scientists have also been able to learn more about Earth's rocks by drilling into its crust. The USA launched one of the first big scientific drilling projects in 1957. It was called Project Mohole. It drilled through the ocean floor off the coast of Mexico, cutting into the sea bed 3.5km below the water's surface. The project dug up rocks over five million years old.

Comparing the interiors of the Earth and the Moon. The Earth's insides are hot and molten, while the Moon is much cooler and solid.

Amazing

Digging deep

The Kola Superdeep Borehole project in northern Russia has drilled down more than 12km, and brought up rock samples more than 2.7 billion years old.

Measuring shakes

Incredibly, scientists have learned about the thickness and **density** of the Earth's crust, mantle and core by studying the shaking movements caused by earthquakes. Earthquakes cause different kinds of shaking motion – some push and pull the crust, others move it up and down. By measuring these motions, scientists can tell how dense the material is beneath the crust, and so build up a picture of inside our planet.

⇧ Open mines, such as this one in Russia, may look very deep, but they only scratch the surface of the Earth's crust.

⇧ A scientific station has been set up on Mount St Helens to measure earthquakes and earth movements.

Rocks

Three main types of rock make up the Earth: igneous rock, sedimentary rock and metamorphic rock.

Igneous rocks

Igneous rocks are formed when molten rock cools and hardens. The igneous rocks formed on the Earth's surface are called volcanic rocks. Basalt is a volcanic rock. Pumice is also a volcanic rock, which cooled as it was blasted out of a volcano. It is rough and full of holes caused by the gas bubbles that frothed in it when it was molten.

Igneous rocks formed beneath the Earth's surface are called plutonic rocks. They cooled down more slowly than volcanic rocks. Granite is a type of plutonic rock.

Project

Rock hunting

Collect local rock samples and try to identify them. Your local library or museum may have information on the types of rock in your area. Perhaps you are living on an ancient volcano, or on ground that used to be part of the sea bed!

⇑ Granite is a plutonic igneous rock, made when molten magma cools deep below the ground. It is extremely hard.

This is a sedimentary rock called a conglomerate, because it is made up of lots of rock fragments cemented together with a finer material. ⇨

Sedimentary rocks

These are the most common type of rock. Sedimentary rocks, such as sandstone, are formed by the breaking down and weathering of other rocks. Material settling on a sea bed can also harden over time to become a sedimentary rock, such as limestone or chalk. These sedimentary rocks are made up of the skeletons of tiny sea creatures. Coal is a sedimentary rock formed from the remains of dead trees and plants.

Metamorphic rocks

Metamorphic rocks are made when one type of rock is changed into a different type by extreme heat and pressure within the Earth's crust (called compression). Igneous and sedimentary rocks can be changed into metamorphic rocks, and existing metamorphic rocks can be changed into different metamorphic rocks.

⇦ Gneiss (say 'neece') is a type of metamorphic rock.

Earth ages

A hundred years ago, geologists studying the Earth knew that most rocks must be extremely old. For example, a rock such as coal – made from ancient trees and the remains of other plants – needs countless thousands of years of compression beneath the Earth's surface to form. However, the geologists could only guess at how old the rocks actually were.

Clocks in rocks

This all changed when scientists discovered 'clocks in rocks'! **Geologists** found they could measure a special property of the rocks – a property called **radioactivity** – that could tell them how old the rocks actually are.

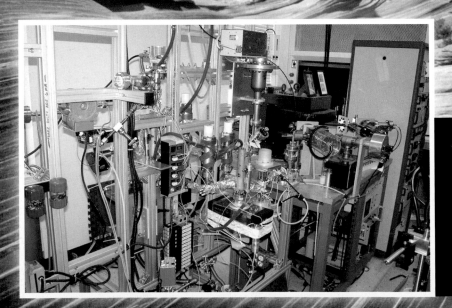

Geologists use simple hammers to collect rock samples. These samples are then examined in the laboratory with complicated instruments, such as this one, which can measure radioactivity to reveal a rock's age.

A timeline

Geologists have named different periods of time in history to help them date rocks. The oldest rocks are about 4 billion years old, from an age called the Precambrian Period. They are metamorphic rocks from north-west Canada. We know more about conditions on Earth in later periods, from the Cambrian Period (about 540 million years ago) through several more periods to the present Quaternary Period, which began about 1.8 million years ago.

Most of the fossils we find are millions of years old. This is the fossil of a fish that lived 56 million years ago.

These ancient sedimentary rocks in Arizona have been worn away by the wind to reveal hundreds of layers.

Fossils

Many sedimentary rocks contain fossils. These are impressions of prehistoric life. Some fossilized creatures look like nothing living on the Earth today, so they must have died out a very long time ago. Fossils of extinct animals added to the evidence that most of the Earth's rocks must be very old indeed.

Water and ice

Three-quarters of the Earth's surface is covered with water. This is made up of six major oceans, along with many seas, lakes and rivers. The Pacific Ocean is the biggest ocean in the world. It is 15 000km wide and 5km deep – that is much bigger than all the land on Earth put together.

Seas and lakes

Earth has many inland seas and lakes, and thousands of rivers. The biggest inland sea is the Mediterranean, which lies between Africa and Europe. In western Asia is the entirely landlocked Caspian Sea, our planet's largest **saltwater** lake. The five Great Lakes in North America are the biggest **freshwater** lakes in the world, containing 20 per cent of Earth's surface freshwater.

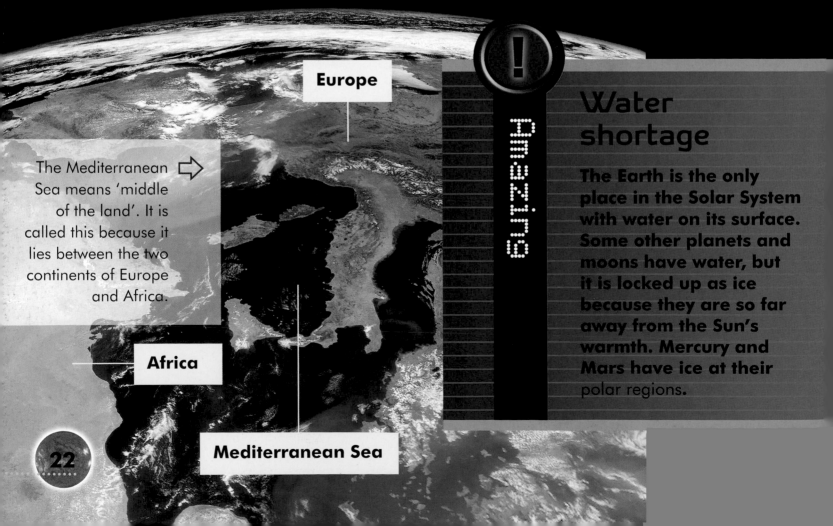

Europe

The Mediterranean Sea means 'middle of the land'. It is called this because it lies between the two continents of Europe and Africa.

Africa

Mediterranean Sea

Amazing

Water shortage

The Earth is the only place in the Solar System with water on its surface. Some other planets and moons have water, but it is locked up as ice because they are so far away from the Sun's warmth. Mercury and Mars have ice at their polar regions.

This is a view of Earth, taken from space. We call Earth the 'blue planet' because most of its surface is made up of water.

In the vast, ice-covered continent of Antarctica the ice is slowly melting.

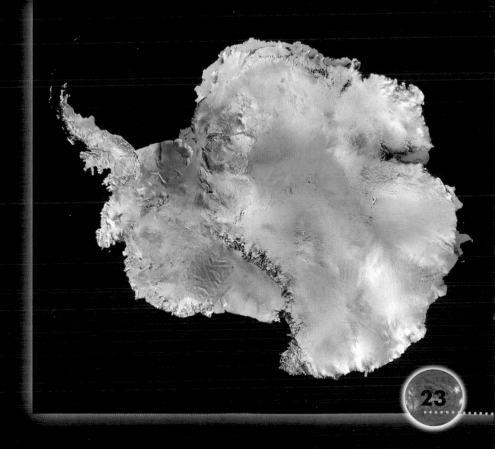

The icy poles

At the top and bottom of the Earth are two areas called the North and South Poles. They are so far from the Sun's rays that they are very cold and covered in ice and snow. There have been many times in Earth's history called **Ice Ages**, when the Earth has cooled so much that the polar ice has spread further, covering more land and making sea levels fall. At the moment, the Earth is warming slightly, so the ice at the poles is slowly melting and sea levels are rising.

The atmosphere

The Earth is surrounded by a protective layer of gases, known as the atmosphere. These gases include nitrogen and also oxygen, the gas all life needs in order to live. The atmosphere contains less of these gases the higher you go. At around 100km high, there is no more atmosphere, and space begins.

A protective shield

The atmosphere is a barrier that protects us from a great deal of harm from space. It absorbs dangerous rays from the Sun. Large **meteoroids** and small comets arriving from space also break up when they enter the atmosphere.

⇧ Are **hurricanes** a result of **global warming**? This picture is of Hurricane Andrew approaching the southern coast of the USA, in August 1992.

A climate blanket

Most of the atmosphere is contained within a layer just 10km deep. Much of our weather takes place inside this layer. Heat and water vapour move around the globe inside the atmosphere, giving us our different weather and climate patterns.

Key Concept

Displays in the atmosphere

Some wonderful sights in the night sky are produced in the atmosphere. Bright flashes of light, called meteors, are caused when tiny bits of comet burn up in the atmosphere. Meteorites are small, harmless chunks of meteoroids. When they travel through the Earth's atmosphere they become spectacular fireballs.

⇧ When small objects from space, such as these meteorites, enter the atmosphere, they burn up and leave a trail behind them.

⇦ The aurorae are multicoloured light displays that take place in the atmosphere. They are caused when particles from the Sun hit the oxygen and nitrogen in the atmosphere, and glow brightly. This is a picture of the Aurora Borealis (Northern Lights), seen over Iceland.

Global warming

Extremes of weather, such as droughts and hurricanes, are sometimes explained as the result of our changing climate, as the Earth warms up. Many people think that this global warming has been caused by human activity. Our cars, planes and factories release too many harmful gases into the atmosphere, damaging it so that it cannot protect us so well. We need to look after our atmosphere so that it can continue to do its job.

Life on Earth

As we have seen, planet Earth is special in many ways. Perhaps its most special quality is that it is a planet upon which many different plants and animals can survive.

⬇ These thick chalk cliffs are made of th skeletons of countle billions of tiny mari lifeforms, which lived more than 10(million years ago.

Life begins

Life first appeared on our planet about 3.5 billion years ago. Simple life was able to change carbon dioxide gas and water into oxygen. Other life forms then developed that made use of this oxygen. About 540 million years ago, at the beginning of the Cambrian Period, the life forms on Earth became more complex and varied, and lots of different plants and animals began to thrive in the sea and on land.

⬆ This is how the Earth may have looked at the beginning of the Cambrian Period, as life on Earth began to flourish.

A suitable environment

There are many different environments on Earth with animals and plants that are particularly suited to them. Life can be found almost everywhere, from blisteringly hot volcanic vents in the depths of the ocean to the freezing world of Antarctica.

⬆ This hot volcanic vent on the sea bed is known as a 'black smoker'. Life may first have developed around these undersea features.

DNA and evolution

All animals and plants contain DNA. **This is a code that sets out how something should look and work. DNA helps all living things to grow and reproduce. If the DNA of an animal or plant changes, it will look and behave differently. This can make it better suited to its environment, and help it to survive. The DNA of life on Earth has been changing for millions of years, which is why so many different types of plants and animals live on our planet. We call these changes** evolution.

Earthwatching

Satellites allow us to study the Earth from space. They have shown us a great deal about its continents, seas, ice sheets and atmosphere. We can also use them to watch any changes that affect our planet.

Mapping satellites

Mapping satellites carefully measure shapes on the Earth's surface, from the peaks of the Himalayas to the valleys of California. They have also mapped the ocean floors in great detail. Some satellites can even show what types of **mineral** are in the Earth's rocks.

At night, satellites show us just how many people live on our planet, when it is lit up from space with bright city lights, road lights and industrial fires.

Orbiting satellites have mapped the contours of the surface of the Earth and the sea bed.

28

Satellite spotting

On a clear night, look up at the sky. If you see a single white point of light, moving slowly in a straight line across the sky, it is probably a satellite. It is orbiting a few hundred kilometres above you. The International Space Station **can appear very bright, as sunlight glints off its large, shiny panels.**

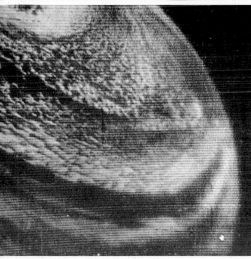

⬆ The first-ever weather satellite image of the Earth was taken in 1960.

Weather satellites

Weather was first monitored from space in 1960. Since then, weather satellites have become very advanced. They can now track clouds and storms, identify rain clouds, see the dust blown out into the atmosphere by volcanoes, and measure the atmosphere's chemical content.

Humans affecting the planet

Tropical rainforests are cleared every year to make way for land upon which farm animals can feed. In 2006, satellites showed that an area of Brazilian rainforest the size of Greece was destroyed in this way. Changes like this harm plant and animal life, damage our climate and could change the way Earth looks for ever. We must all take better care of Earth so that it continues to be one of the most beautiful and life-filled planets in our Solar System.

Glossary

asteroid   a lump of rock in space

Asteroid Belt   a band of space between the planets Mars and Jupiter containing thousands of large asteroids

climate   the average temperature and weather experienced in different parts of the world

comet   a huge ball of ice and dust in space

continent   a large landmass. There are seven continents on Earth: Asia (the largest), Africa, North America, South America, Antarctica, Europe and Australia (the smallest)

continental drift   the movement of the Earth's continents in relation to each other

core   the heaviest, thickest part of a planet, lying at its centre

crater   a large, bowl-shaped pit which has been blasted out of a solid surface by the impact of an asteroid. Volcanoes also have small craters at their tops

density   how heavy an object is compared with its volume

DNA   the substance in the cells of all plant and animal life on Earth. DNA stands for deoxyribonucleic acid

earthquake   a shaking motion in the Earth's crust caused when two plates slide past each other

evolution   the process by which life forms on Earth have changed to survive in their own particular environment

fossil   the preserved remains of prehistoric plants and animals in rock

freshwater   water that does not have much salt. Freshwater is found in most lakes and rivers

geologist   someone who studies the structure of the Earth's crust and its layers

global warming   a gradual rise in the average temperature on Earth

gravity   a force that acts throughout the Universe. The Earth's gravity holds you to its surface, and the Sun's gravity holds the Earth in its orbit. The bigger the object, the more gravity it has

hurricane   a powerful storm of rain clouds hundreds of kilometres wide, with average windspeeds over 100kph

Ice Ages periods of global cooling lasting around 100 000 years, during which the polar ice caps grow

ice caps vast, thick sheets of ice covering the Earth's polar regions

International Space Station a large structure orbiting Earth in which research about space takes place

lava hot, molten rock which has bubbled up from below a planet's crust

magma hot, molten rock that is beneath the Earth's crust

mantle a layer of molten rock between the Earth's crust and its core, about 2900km thick

meteoroid a small rock in space, usually a chip off an asteroid

mineral a solid material made of chemicals

molten something so hot that it is in a melted state

Moon the Earth's only natural satellite. Other natural satellites are also known as moons (with a small 'm')

orbit the curved path of a planet or another object round a star, or a moon round a planet

planet a large, round object orbiting a star

plate a large segment of the Earth's solid crust

polar regions the cold areas at the very top and bottom of a planet. Polar regions are cold because they are the areas of a planet that are furthest from the Sun

radioactivity energy given off by rocks which helps scientists to work out how old the rocks are

saltwater water that contains a lot of salt. Saltwater is most often found in seas and oceans

satellite a small object in orbit around a larger object. Man-made satellites are sent up into space to study the Earth

Solar System our part of space, containing the Sun, the planets and their moons, asteroids and comets

space everything outside the Earth's atmosphere

Sun our nearest star, a huge ball of burning gas

Universe everything there is, to the unimaginably distant reaches of space

volcano a mountain built up by the eruption of hot, molten rock and piles of ash

weathering the wearing away of Earth's surface by weather, such as rain and wind

Index